Planetary Defense

Patrick H. Stakem
2018, 2022

Number 25 in Space Series

Table of Contents

Introduction..4
Author..5
A note on Units..5
What do we have to be afraid of?...6
    Near Earth Objects...6
    Potentially Hazardous Asteroids..8
    Solar Storms..8
    Cosmic Rays..9
    Coronal Mass Ejections...9
    Space Debris...11
    Zombie Sats...13
    Satellite on-orbit collision..14
    Booster debris...15
Space Station Mir..15
Chinese Space Station uncontrolled re-entry............................16
The LDEF Mission..16
    What we did with Apollo hardware......................................17
    Meteor Airburst...18
Notable events..20
    The Chicxulub crater...20
    The Tunguska event in Siberia..21
    The Carancas event in Peru...21
    The Sutter's Mill meteorite..22
    Some other events...22
What are we doing about planetary defense?............................23
NASA/JPL's Sentry Program...25
NASA's Planetary Defense Coordination Office........................26
Asteroid impact avoidance...26
Non-Near Earth Issues..27
NASA Orbital Debris Program Office.......................................27
    RemoveDebris..29
    WISE...30
NASA's Asteroid Redirect Mission..31
    NEAR-Shoemaker...32
    Deep Impact Mission..32

Testing Earth's Planetary Defense..32
    The Martian Planetary Defense Force...............................33
Launch Vehicles...33
DoD's Role, NORAD...33
Other Spacefaring Nation's approaches and cooperation. 34
Project Icarus...36
Wrap-up..36
Glossary of Terms..37
Bibliography...42
Resources...45
If you enjoyed this book, you might also be interested in some of these..47

Introduction

This book covers the topic of Earth's Planetary Defense. This is not for the purposes of repelling an alien invasion, although the military has probably thought of that. Any civilization that could reach us is undeniably more technological advanced.

We are focused, however, on big chunks of rock that could cause massive damage, and end civilization as we know it. It has happened – ask the dinosaurs. Ocean impacts, and Earth is covered by 75% water, recall, could trigger massive Tsunamis.

The U. S. National Science and Technology Council is on record as saying we are unprepared as a nation for a large asteroid impact event. The last impact of a object some 10 km in size, 66 million years ago, caused an "extinction event."

A four meter asteroid enters the Earth's atmosphere about

once a year. For seven meters, its 5 years. That size produces a Hiroshima-sized explosion. For 20 meter asteroids, its twice per century. We just had one in 2013. We have had a success rate of four asteroids detected before they hit Earth's surface. It's the ones we don't detect that should trouble us. That, and the fact we don't have a clue on what to do about it, if its detected.

About 100 tons of stuff enters the atmosphere every day as dust (and settles on my car...). On average, we get about 30 entries of smaller items, most of which burn before they hit the ground,

The non-profit B612 Foundation said, in early 2018: "It's a 100 per cent certain we'll be hit by a devastating asteroid, but we're not 100 per cent sure when."

What do we have to defend against? A wide variety of Near-Earth Objects, a bunch of space debris in orbit that we are responsible for, radiation blasts and chunks of the Sun, lot of stuff out there that gets caught up in Earth's gravity, or that we cross paths with. We are better than we were at detection. It remains to be shown if we are any good at deflection. It's high stakes.

Let's make up a simple example. On any give day, you might have a small rock hit the windshield of your car. This might happen several times a year. Most of them bounce off. Every once in a while, one will cause a pit or a crack. If you're driving behind a gravel truck, you could have multiple hits. If you drive under the wrong bridge at the wrong time, and a big rock lets loose...

Author

The author has a BSEE in Electrical Engineering from Carnegie-Mellon University, and Masters Degrees in Applied Physics and Computer Science from the Johns Hopkins University. During a career as a NASA support contractor from 1971 to 2013, he worked at all of the NASA Centers. He served as a mentor for the NASA/GSFC Summer Robotics Engineering Boot Camp at GSFC He taught Embedded Systems for the Johns Hopkins University, Engineering for Professionals Program, for the Graduate Computer Science Department, and for Capitol Technology University. He has done several summer Cubesat Programs at the undergraduate and graduate level.

A note on Units

I am fairly conversant in both English and Metric units (what is the metric equivalent of furlongs per fortnight?). Metric (SI) is mandated for NASA usage now, for interchangeability with our partner space faring nations. When a lot of the legacy flights discussed here were flown, English units were the norm. I have tried to keep the units comparable to the mission at the time. Conversions are easy enough, but units conversion is a source of error. It's not what you know about units and measurement, its how you think. And, I still think English units (even the English use Metric now), and convert in my head or on my phone.

For scientific/engineering work, the Metric system is well thought out. For artisans, the English system served well, as most units were divisible by 2. Which is easy. Fold the cloth. Hopefully, when we are all taught Metric first, some one will still remember the conversions. You just need a good slide rule....

What do we have to be afraid of?

We are not talking about defending our planet against alien invasion. There are worse problems. Solar storms threaten our space, air, and land assets. Large asteroids hit the surface and explode with the power of atomic weapons. One of these could be mistaken for a nuclear weapon, and cause a deadly exchange. We don't need aliens to obliterate us, we can do that ourselves.

Near Earth Objects

An NEO is a solar system object whose closest approach to the Sun is 1.3 AU, and that comes in close proximity to the Earth There are 15,000 known asteroids in this category, 100 comets, some solar orbiting spacecraft, and meteoroids. All these have the potential of striking the Earth. They are closely tracked from the ground, by NASA's Planetary Defense Coordination Office. A joint US/EU project called the Spaceguard Program is tracking NEO's larger than 30 meters. Three NEO's have been visited by spacecraft. NEA's are grouped into one of four categories, based on their perihelion, aphelion, and semi-major axis.

There are four groups of NEO, named for the first

member observed. Group Atira has members whose orbits are contained within the orbit of the Earth around the Sun. Group Aten are Earth-crossing, with a semimajor axis smaller than Earth's. Group Amor is between the Earth and Mars. Group Apollo asteroids have a semimajor axis larger than Earth's.

If NEO's enter the atmosphere, they heat up and burn. Sometimes, enough is left to hit the ground, appearing as a rock. The easiest place to find meteors (as we call asteroids that hit the ground) is Antarctica, where they stand out against the snow and ice. A lot of the Antarctic meteors come from Mars, as one of my old professors proved.

Most of the asteroids are found in the main asteroid belt, between Mars and Jupiter. A lot of them hang out with the big planets, Jupiter, Saturn, Uranus, and Neptune. If they orbit the Sun, they are asteroids; if they are captured by a planet's gravity, they are moons.

In the main asteroid belt, about half the mass is contributed by 4 asteroids, Ceres, Pallas, Vesta, and Hygiea The total mass of the belt is estimated to be about 4% of Earth's moon. An intruder in the asteroid belt is the dwarf planet Ceres, topping the scales at 950 km diameter. There are hundreds of thousands or millions of objects in the belt, some being dust particles. Still the asteroid belt is mostly empty, routinely passed through by spacecraft on their way to the outer planets.

In 1807, Juno and Vesta were discovered. By 1868, the

number of asteroids discovered reached 100. The advent of the use of photography in astronomy opened the floodgates of new discovery.

Asteroids are considered in three classes, silicate, metal-rich, and carbonaceous.

Occasionally, our solar system get a visitor from afar, usually a comet, which makes a loop around the sun, and heads back out in space. Generally, our solar system bodies orbit the Sun in a disk called the ecliptic plane. Comets that are not necessarily in orbit around our Sun can take a path that are highly inclined to that plane. The first asteroid at a very high angle with respect to the ecliptic was recently observed. That means it did not originate in our solar system, but came from some where else in our galaxy, or beyond. There is no particular name for this class of objects, but the title "Exeroid" has been suggested. This will have to be cleared with the International Astronomical Union. The object was observed by a telescope on the Hawaiian mountain of Hakeakala. It is the first time an interstellar object has been observed. It was named Oumuamua, in Hawaiian, "messenger from a far, arriving first." It came from the direction of the constellation Lyra.

The Minor Planet Center collects all the world wide observational data for minor planets (asteroids and comets) and maintains a database. It is located at the Smithsonian Astrophysical Observatory, at Harvard University. It maintains an NEO website and blog. At this writing it has a total of 524,000 minor planets, attributed to more than a thousand astronomers at more than 200 institutions.

Potentially Hazardous Asteroids

A PHA is defined by its potential to come close enough to Earth to pose a threat. This is determined by an Earth minimum orbit insertion distance of 0.5 AU or less (around 7,500,000 km) and an absolute magnitude of 22 or less. What's the significance of the magnitude? We assume an albedo of 14%, which would give a diameter of 140 meters or so. Usually the asteroid can't be imaged or resolved in the telescope.

For most of the known asteroids, not much is know about their composition or even their orbit. The mass is estimated from the diameter, but this depends o the composition. Density is obtained by "enlightened guesswork." For irregular objects, not spherical, the surface gravity will vary by location.

Asteroids were predicted by Kepler between Mars and Jupiter. He never observed one, but he based his prediction on the large gap between those planets.

Solar Storms

The Earth and other planets are constantly immersed in the solar wind, a flow of hot plasma emitted by the Sun in all directions, a result of the two-million-degree heat of the Sun's outermost layer, the Corona. The solar wind usually reaches Earth with a velocity around 400 km/s, with a density around 5 ions/cm^3. During magnetic storms on the Sun, flows can be several times faster, and stronger. The Sun has an eleven year cycle of maxima. A solar flare is a large explosion in the Sun's atmosphere that can release as much as 6×10^{25} joules of energy in one event, equal to about one sixth of the Sun's total energy output every second. Solar flares are frequently coincident with sun spots. Solar flares, being releases of large amounts of energy, can trigger Coronal Mass Ejections, and accelerate lighter particles like protons to near the speed of light.

We get all our energy from this huge thermo-nuclear reactor some 8 light-minutes away. As you get closer to the Sun, its gets hotter, and there is more radiation in terms of energetic particles.

The March 1989 severe geomagnetic storm blacked out Quebec for some 9 hours. It was caused by a solar coronal mass ejection, that hit the Earth 3 days later. The charged particles induced electric currents in the ground. On-orbit satellites were affected. Currents were induced in very long electrical transmission lines, such as those of Quebec's power grid. Radio communications were also adversely affected.

Cosmic Rays

Galactic Cosmic rays are actually heavy ions, not originating in our solar system. The actual origin is unknown. They carry massive amounts of energy, up into the billions (10^9) of electron volts.

Coronal Mass Ejections

A large solar flare occurred in September of 1859, and was observed by British amateur astronomer R. C. Carrington in his private observatory on his estate outside of London. Both the associated sunspots and the flare were visible to the naked eye. The resulting geomagnetic storm was recorded by a magnetograph in Britain as well. They also recorded a perturbation in the Earth's ionosphere, that we now know is caused by ionizing x-rays. In 1859, this was all observed, but not understood. Even the ionosphere was not known to exist at the time. Now, we know a Coronal Mass Ejection from the sun, associated with a solar storm, is first seen as an energy burst hitting the Earth, and later by vast streams of charged particles, that travel slower than the speed of light. At normal levels, these particles are seen as the Northern or Southern lights. The Earth's magnetic field is affected.

What did happen, and was not immediately associated with the solar storm, was interference with the early telegraph systems of the time. The telegraph was relatively new, and wires stretched for many miles. Think of them as long antennas. The telegraph equipment was damaged, and large arc's of electricity started fires and shocked operators. No fatalities were reported. The employees of American Telegraph Company in New York found they could transmit messages with the batteries of their systems disconnected. The Northern lights were visible from Cuba. This was the largest such solar flare in at least 500 years...and so far.

What if such a super flare occurred today? First, we would have warning from sentinel satellites such as the Solar Dynamics Observatory, that are closer to the sun, and detect the passage of particles. They can tell us about this via radio, which travels faster than the particles. So, we would have a day or so's notice. All of our modern high-technology infrastructure would be at risk of damage, from the electrical grid to the Internet. Most of our satellites would be damaged, removing services we rely on such as long distance data communication, and navigation. It would be much better to turn everything off, and ride out the storm. Even that might not prevent major damage to networks. When is the next large solar event? Even the Astrophysicists can't tell us that. Only that it will eventually occur. We need to know more about the Sun, which is the focus of the recently launched Parker Solar Probe.

Space Debris

Sometimes we are out own worst enemies. Orbital debris can come from many sources. It can be naturally occurring rocks and captured meteoroids. It can come from satellites collisions and explosions. Liquid fuel, from a dead satellite or leakage, freezes in space and provides yet more debris problems. There have been 5 known collisions of satellites in orbit so far.

All of the junk, down to the size of bolts, is tracked by the U.S. Air Force. They know of 18,000 objects in orbit, of which 1,400 are operational satellites. A good job for a robotic servicer in LEO would be to collect the trash, put it in a canister, and kick it off to re-enter and burn in the atmosphere. There are estimated to be 170 million chunks, smaller than a centimeter, any of which can ruin your mission.

Just as I was writing this, a chunk of the defunct Iridium-70 satellite hit California.

Known space debris includes Astronaut Ed White's outer glove, lost on his space walk; Michael Collins's camera from Gemini-10; a wrench, pair of pliers, and a tooth brush; and a complete tool bag from an STS-26 EVA.

In addition to our own junk, space itself provides dust particles, up to and including asteroids capable of ending all life on Earth with a collision.

In normal operations, the Earth's magnetic field deflects charged particles from the station. Energetic space particles may pass through the ISS with negligible effect. On rare occasion, the station needs to do a damage avoidance maneuver to avoid a collision. To date, no evacuation of the Station has been necessary. It does a debris avoidance maneuver if there is a greater than 1 in 10,000 chance of a strike. This tends to happen about once per year.

It the warning comes too late for maneuvers (usually a few days), the crew shelters in the attached return capsule.

For mitigation, the ISS uses Whipple shielding, named for its inventor. This uses a thin outer bumper, spaced away from the hull. The idea is, the bumper breaks up the debris so that penetration doesn't happen. It works most of the time. Think of it as the ISS's bullet proof vest. It works best with a foam material behind it.

In 2009, damage was noted on a radiator assembly, that had to be taken off line. The radiators contain ammonia. Astronauts on EVA have reported sharp edges on handholds due to small impacts. Some of the windows show obvious effects of particle damage, just like the windshield on your new car when you drove behind the gravel truck. The solar arrays, big targets, have numerous punctures, but they are not showing degradation from this. So far, most damage is on the leading edge, along the velocity vector, but particles can come from any direction. A working return capsule for the astronauts is kept docked to the station at all times, as a lifeboat. The amount of time a capsule can be docked is limited, and a new upcoming crew brings a new capsule, and the downgoing crew uses the old one.

There have been on-orbit tests of anti-satellite weapons, by the U.S., Russia, and China which resulted in a large amount of debris. In addition, failed on-orbit rendezvous usually involved a low speed collision, and more debris creation. There was the 1996 collision between the French Cerise military reconnaissance satellite and debris from the Ariane rocket. There was the 2013 collision between debris from the Chinese Fungyun FY-1C satellite and the Russian BLITS nano-satellite. We should also mention the 2013 collision between two CubeSats, Ecuador's NEE-01 Pegaso and Argentina's CubeBug-1, and the particles of a debris cloud around a Russian Tsyklon-3 upper stage from the launch of Kosmos 1666. A 1985 U.S. Anti-satellite test created thousands pieces of debris greater than 1 cm. Most of this reentered the atmosphere within 10 years.

China is responsible for the largest single space debris incident resulting from a 2007 test of anti-satellite weapons. It transformed one satellite into nearly 2,500 pieces, golf-ball size or larger. The test was conducted in an orbital zone where most near-Earth missions are located. Some satellites have had to maneuver to avoid collisions.

NASA says that at least one piece of space debris falls to Earth daily, and has for the past 50 years. There have been injuries, such as the Japanese fishermen hit in 1969. A lady in Oklahoma was hit by a piece of debris in 1997 but was not injured. This turned out to be a pieces of a Delta-II rocket propellant tank.

An early attempt (1962) to characterize the space meteorite environment, prior to the manned Apollo missions, was the Pegasus project. This was a spacecraft launched to orbit by a Saturn-I rocket. It unfolded large wings that detected and reported strikes. There we three Pegasus flights.

The Pegasus payload was a 3,200 pound satellite, with 95 foot wing panels that folded out from the body. The total area of the wings was 2,300 square feet. With the right sun angle, the Pegasus could be seen by the unaided eye, from the ground. There were 116 hit detectors, shielded by panels of varying thicknesses. There were also vibration and acoustic sensors.

The mass range of the micro-meteroids was 10^{-7} to 10^{-4} grams. Pegasus also monitored the on-orbit radiation environment. From the Pegasus data, it was decided that the flights to the moon and back represented a "reasonable risk."

Zombie Sats

Zombie sats are non-functioning satellites in orbit. They may have experienced a failure, and are no longer functional. They remain in the same slowly-decaying orbit, however. The Intelsat Galaxy-15 is an example. It was in geostationary orbit when the ground lost control, and it began to drift. There was a potential of collision with other, operating satellites. Later, control was recovered, and it was directed back to its correct orbital position.

The first U. S. satellite, Vanguard-1, and its upper stage are still in-orbit. It was launched in 1958. The Soviet-era RORSAT, with its BES-5 nuclear reactor, is still in orbit. In 2015, the USAF Defense Meteorological Satellite (Military equivalent of the GOES weather satellite) exploded in orbit, creating some 150 chunks of satellite in orbit.

Of the 7,000 or so satellites placed into Earth orbit so far, 1,500 are still functioning. The rest are zombie-Sats.

Satellite on-orbit collision

A collision between two satellites occurred in February of 2009. One was a Russian Strela-class military satellite, massing 950 kilograms. The other was the commercial Iridium-33 communications satellite. What was the cause? They were in the same place at the same time. The Russian spacecraft had been deactivated, and was classified as space debris. The Iridium was operational, but was destroyed. This produced some 2,100 pieces of debris. There is a catastrophic satellite collision every 5-9 years.

And, the bad news is, the collision created a thousand pieces of space debris larger than 4 inches, and many more smaller ones. In March 2012, a piece of the KOSMOS 2251 passed by the International Space Station, prompting the crew to take refuge in the attached Soyuz return craft as a precaution. The ISS frequently does obstacle avoidance maneuvers.

It has been necessary to occasionally conduct Debris Avoidance Maneuvers, to avoid known space debris nearing the station. It has been hit many times, but the hull has not been breached.

The SNAP-10 spacecraft was launched in 1965. It included a nuclear fission device for electrical power. It operated for 45 days before an electrical failure. It was placed in a 1,300 km orbit, where it is expected to remain for around 4,000 years. It was noticed that in 1979, the spacecraft was shedding traceable debris. It is not known if radioactive material is included.

There are actually more than 30 small nuclear reactors in orbit, mostly Soviet.

The liability for space debris is defined by the United Nations Convention on International Liability for Damage Caused By Space Objects. The nation (or nations) that launched the object is "absolutely liable" for compensating the injured party(s), according to the convention agreed to by most space faring nations.

Booster debris

After the first stage of the launch vehicle has done its job, it usually crashes into a remote part of the ocean. The next stage may also. The final stage usually accompanies the payload into orbit. The Shuttles' boosters were dropped in the Atlantic, recovered, refurbished, and reused. The large liquid fuel tank was crashed into a remote part of the ocean. These are a particular problem, because they always have residual fuel, which decomposes.

A Chinese Long March-4 upper stage exploded in orbit in 2000, creating a vast debris cloud. A Russian booster exploded in 2007, captured in images. The debris cloud was tracked on radar. More than 1,000 fragments were identified.

In 2001, part of a U. S. upper stage of the NavStar-32

satellite crashed into the Saudi Arabian desert.

Space Station Mir

Quite a lot of lessons were learned from the impact of small debris on the Soviet Space Station Mir. The Environmental Effects payload was returned and studied, adding to knowledge applied to the ISS. It was in orbit for 3,644 days before reentering the atmosphere.

Chinese Space Station uncontrolled re-entry

The out-of-control Chinese Tiangong-1 Space Station was going to hit the Earth. It is massive enough to cause extensive damage if it goes anywhere except the satellite graveyard in the South Pacific.

The point in the Ocean farthest from land is called the pole of inaccessibility. This is the best point to aim for, if you're de-orbiting something large. It happens to be located in the South Pacific, some 1,600 miles south of the Pitcairn Islands. There are some 260 satellites on the ocean floor at that point. This will be an interesting place for future archaeologists. Among other things, the 120-ton MIR space station is there. It's entry into the water was observed by fishermen. This is also where supply modules from the ISS, loaded with trash, are sent. Sometimes, things don't quite work out. 36 tons of a Russian Salyut Space Station came down on land in South America.

Just one known person, Lottie Williams of Tulsa, Oklahoma, has ever been hit by space junk, according to news reports, but she wasn't hurt. On Jan. 22, 1997, Williams saw "a flash of light resembling a meteor." A few moments later, something metallic fell onto her shoulder. NASA said her incident came close to the timing of the re-entry and breakup of a second stage of a Delta rocket coming into Earth's atmosphere.

The LDEF Mission

The Long Duration Exposure Mission was placed in a 275 mile orbit by Space Shuttle Challenger in 1984. It spent 69 months in space, collecting information on long term exposure of materials to the space environment, and the meteoroid environment. It was to be periodically returned to Earth, refurbished, and re-orbited, but instead, the original mission was extended. It was recovered and returned to Earth by Shuttle Columbia. Although it was expected that meteoroid impacts would be the major phenomena, over 30% of the observed impacts were from debris. This lead to the realization that the debris environment was larger than suspected. In some cases, the impacting body was still embedded in LDEF's structure. There was penetration of up to 40 mils of aluminum observed. In total, 34,000 impacts were cataloged, the largest being 0.57 cm.

Characterizing the "space dust" environment, it is estimated that over 14 million tons fall to the Earth's surface each year. The number of meteors entering the atmosphere is estimated to be between 10 and 20,000 tons. Most of these burn in the atmosphere, but when they do hit, they can be catastrophic. Generally, if the chunk is big enough to do damage, it can be tracked.

Sometimes you miss them. Like in the Urals in Russia in February, 2013. This injured hundreds, but did not hit anything in the town. There could have been hundreds of deaths.

Impact events are not limited to Earth, but those are the ones that concern us most. Look at the full moon sometime – those are all impact craters. Since our moon has no atmosphere, the asteroids don't burn up before impact. In 1992, an impact was directly observed on Jupiter, involving the comet Shoemaker-Levy-9. It had been captured by Jupiter, and was in orbit around that planet. Its demise was observed by the Galileo spacecraft, then heading to Jupiter. The impact was on the side of Jupiter opposite the Earth in 1994. It was also observed by the Ulysses and Voyager-2 spacecraft. The impact resulted in visible fireballs, and auroral emissions. There was a new dark spot observed on Jupiter, more than 7,000 miles across, leaving a visible scar. The energy release was estimated to be 6 million megatons (of TNT). This kicked off the expansion and increased funding of asteroid detection programs across the globe.

What we did with Apollo hardware.

The first and second stages of the Saturn-V vehicle were splashed into the ocean.

The third stages of the Saturn-V were deliberately put into a solar orbit, so as not to endanger current and future missions.

The bottom part of the lunar lander stayed on the surface, and they are still there. Most of the crewed portion of the

LEM's were deliberately crashed into the lunar surface, to provide data for the seismic instruments left on the surface. The Lunar Modules *Antares, Falcon, and Challenger* impacted the lunar surface.

On the ill-fated Apollo-13 mission, the lunar module *Aquarius*, serving as the life boat on the return trip, was reentered into the atmosphere in a controlled manner, and burned during reentry.

Meteor Airburst

A meteor entering the Earth's atmosphere can cause an explosion. These may have started out as asteroids in space. The extreme example is the 1908 Tunguska Event in Siberia, an explosion equal to an atomic weapon. n the 500 kilotons area.

Asteroids of up to 4 meters in size impact the Earth about once per year, releasing the equivalent of 3 kilotons of energy. Asteroids of up to 70 meters impact every 1900 years or so, with an energy of 16 megatons (TNT equivalent). These can trigger extinction events. The one that caught the dinosaurs off guard was probably 10 kilometers in diameter. In 1949, the same region was treated to a meteor shower, with no damage reported.

An extinction event is defined as "a widespread and rapid decrease in the amount of life on Earth." Given that over 98% of the species we know of are extinct, we need to be careful with the home planet. There have been several mass extinctions on Earth since life began, none because of us. So far.

There was a documented case in 1490 in China, *the Ch'ing-yang even*t, that resulted in 10,000 deaths. It was an extremely bright bolide, that usually explodes in the atmosphere. It can also result in a crater-forming event.

The Earth Impact Effects Program characterizes the average energy of airbursts, and the average frequency of the event. This ranges from a 4 meter, 0.7 kiloton airbusrt every every 1.4 years, to a 70 meter, 15 megaton event, every 1900 years.

The Chelyabinsk Meteor airburst occurred shortly after dawn in February 15, 2013. That city in Russia is located near the border of Europe and Asia. The meteor came in at 34,000 miles per hour over the Ural Mountains, and exploded somewhere around 16 miles from the surface. The height of the explosion limited damages to the town, which were extensive enough as it was.

Chelyabinsk meteor event was widely photographed, and went viral on the internet. The superbolide was estimated to be 20 meters in size. The air burst occurred around 30 kilometers altitude, causing a large shock wave, and scattering small fragments over a large area.

In this event, some 1,500 people were injured enough to seek medical help. More than 7,000 buildings in the area were damaged, mostly by broken windows. The bolide was estimated to mass some 12,000 metric tons, and measure about 20 meters in diameter. It was a total surprise, and had not been observed or tracked before it exploded. The explosion was detected by monitoring stations of the Nuclear Test ban organization. The sound waves went around the planet several times. A weather satellite caught a picture of the meteor entering the atmosphere, and it was found later.

The immediate effect was a flash as bright as the sun, and a shock wave that injured over a thousand, and blew out windows for miles. The visitor was estimated at some 50 feet in diameter, massing around 10,000 tons. The effective power of the explosion was 500 kilotons of TNT.

We know more about meteor airbursts now, with increased monitoring under the Comprehensive Nuclear-Test-Ban treaty, using improved detection technology. In 2017, there were 26 events noted.

Barringer, or Meteor Crater, is located northeast of Flagstaff, Arizona, a short drive. I've been there, and it is impressive. Puts the fear of meteorites into you. It was originally thought to be the result of a volcanic eruption, but was recognized in 1905 as an impact crater, and named after the guy who figured it out, Daniel Barriger. The site is now privately owned, and has a visitor's center.

The crater is some 1,200 feet in diameter, with a rim rising 150 feet above the rather flat landscape. The center has some 700 feet of rubble. The impact is thought to have been some 50,000 years ago. At that time, the location was grasslands. It must have really shocked the locals, the woolly mammoths. It was probably known to the local native Americans.

Finding meteorites on Earth allow researches to understand their composition, and perhaps determine their origin. Since the Earth is ¾ covered by water, we are only seeing about ¼ of those that reach zero altitude. There is currently an ongoing student project in Lake Michigan to recover meteorites from underwater.

Notable events

This section discusses some of the dangerous events from space that have happened in the past. The most notable is the extinction event in the time of the dinosaurs.

The Chicxulub crater

There is an impact crater on the Yucatan Peninsula, caused by a body some 10-15 km in diameter. It is thought to have caused global climate disruption, leading to the extinction of some 75 percent of the plant and animal life on Earth. The crater itself is more than 100 miles in diameter. Some of the crater is located on land, with the rest on the ocean floor. Samples from the impact are dated as 66 million years old. Estimated energy release from the impact was the equivalent of 10^9 atom bombs. The dust caused by the impact covered the entire surface of the Earth. The atmosphere had increased carbon dioxide for decades, resulting in a mini-greenhouse effect.

The Tunguska event in Siberia

If the Earth is going to get smacked by a large meteor, Siberia is a good location. It is sparsely populated and isolated. This event happened in 1908. Two square kilometers of forest was flattened, knocking over an estimated 80 million trees.. It was the largest impact in recorded history. It is not known if our visitor was a meteor or a comet. No impact crater has been found. Based on the damage, the strike was the equivalent of somewhere in the range of 10 to 30 megatons of TNT. There were eye witnesses, but no human fatalities are known. Seismometers as far away as the U.S. registered the impact.

It took ten years to get a proper expedition to the remote site. Over the next ten years, three more expeditions were mounted. Eventually, an aerial photographic survey was conducted. Analysis and discovery continues to this writing, including 3-D computer modeling, and ground penetrating radar.

This is THE event that got people thinking about avoidance of impacts from space, well before the space age.

The Carancas event in Peru

Peru got smacked with a meteorite in September of 2007, near the village of Carancas in the southern part of the country, at an altitude above 12,000 feet. It created a crater which spewed noxious gases. Anyone who approached the crater got sick from these emissions. This included some 600 local villagers. This was later suspected to be caused by arsenic poisoning. The local ground water is known to contain arsenic, and a lot of the ground water is the crater was vaporized into the atmosphere.

The crater was some 15 feet deep, 43 feet in diameter. The impact shock shattered windows in the village. There were no causalities or injuries from the impact. Initial assumptions was that the country was under attack from its neighbor, Chile.

This event could have been much more tragic if the meteor had impacted in the village. There are eleven known impact craters in South America.

The Sutter's Mill meteorite

This event occurred in 2012, in the area of the California gold fields. Some of the oldest material in the solar system was discovered in the debris. There were small diamond grains.

A large bolide entered the atmosphere, and resulted in an air burst. The resulting meteorite debris were tracked by weather radar. In addition, the event was recorded by two infrasound monitoring stations of the Comprehensive Nuclear-Test-Ban Treaty Organization's International Monitoring System.

The meteorite was estimated to have been 2-4 meters in

diameter at entry, and resulting in a 4-kiloton explosion at around 48 kilometers altitude.

Most debris recovered was on the order of grams. A rainstorm washed away a lot of the material before it could be collected.

Some other events

In 1969, a large meteorite near Murchison, Australia. It was rich in organic compounds. The collected material from the meteor massed 100 kg. Fifteen amino acids were found in the collected debris.

The Eltanin asteroid hit the Pacific Ocean about 2.5 million years ago. It is estimated to be in the range of 1-4 km. There remains evidence of a huge Tsunami. It was calculated that a wall of water 200 meters high struck present-day Chile. It is possible that such an event lead to the Biblical story of the Deluge, which occurs in various books of the world's religions. This event described is probably not a myth.

There are 130 potential impact craters that have been found around the globe.

What are we doing about planetary defense?

U.S. National Space Policy directs the NASA Administrator to:

"pursue capabilities, in cooperation with other departments, agencies, and commercial partners to detect, track, catalog, and characterize near-Earth objects to reduce the risk of harm to humans from an unexpected impact on our planet."

Planetary defense is being addressed at the highest levels of the U.S. Government. The National Science and Technology Council is part of the Executive branch. Under this is the Office of Science and Technology Policy, charged with providing advice to the Executive Office. There has been established A Damien Inter-Agency Working Group, where Damien is an acronym of "Detecting and Mitigating the Impact of Earth-Bound Near-Earth Objects." In government, it is important to have an awesome acronym.

The National Near-Earth Object Preparedness Strategy and Action Plan is a document that address the topic of Planetary Defense over a ten year span. It is used to co-ordinate efforts among government agencies. The plan has five major goals. The first is to focus on NEO detection, tracking, and characterization. The second goal is to improve models and prediction, and to address Information integration for decision making. The third goal is to develop technologies for NEO deflection and disruption. NASA is the lead for this effort. The fourth goal is to increase International Cooperation on NEO Preparedness. This will involve International support and education. The fifth goal is to "strengthen and routinely exercise NEO Impact Emergency Procedures and Action Protocols."

Impact prediction is a critical process to develop, because we any not be able to prevent the strike, but we might still get the target area evacuated. The first successful prediction was made in 2008, for an asteroid that hit Sudan, by Pan-STARRS, a pair of 1.8 meter telescopes in Hawaii.

The first step in the process is to collect sky survey data from multiple sources and points-of-view. Two major observatories are located in Hawaii, with two more in the southwestern United States. All the data needs to go into a database, so it can be correlated with other observations of the same target, and the data needs to be available on the web for analysis at different centers. It's not quite there yet. Over the last decade, there have been 3 predicted impacts, as opposed to over 500 unpredicted.

Both ground-based and space based assets are used. Ground based telescopes do better at night, and are useless for asteroids coming from the sun's direction.

NASA has a Planetary Defense Officer, working with the Damien Group. Other members are from NIST, NSF, State Department, Homeland Security, USAF, the Lawrence Livermore National Laboratory, and others. Un-predicted asteroid strikes are noted by the in-place infra-sound array, intended to detect nuclear detonations. Space based assets can observed in all frequency bands, including those that are blocked by the atmosphere. So far, the United States has been responsible for around 95% of all near Earth objects found. Listen up – these can hit anywhere on the globe.

NASA's latest asteroid alert system provides us up to a 5 day warning. It uses a ground-based facility, a telescope in Hawaii. If it detects something that could be a problem, it notifies other observatories. From multiple observations from different locations, an orbit can be calculated. About 5 new asteroids are found, every night.

The Kitt Peak observatory in Arizona hosts Spacewatch, a survey of minor planets This approach was tested with asteroid 2016 UR36. It was judged not a credible threat, due to the fact that it would pass Earth beyond the moon's orbit.

The Large Synoptic Survey Telescope is under construction as of this writing, and will have an 8.4 meter primary mirror, and a 32 gigapixel camera. It will continually photograph the entire night sky. It will be located in Northern Chile at 2,680 meters. It will join the existing Gemini South and Southern Astrophysical Research Telescopes. It was started with private funds, but was taken over by the National Science Foundation. The plan is to make all data available to any one who wishes it, shortly after it is compiled.

The data is expected to include over 200,000 pictures a year, and reach 1.28 petabytes in size. This will require automated processing by large, probably GPU-based arrays.

NASA/JPL's Sentry Program

The Sentry Program produces and analyzes the Sentry

Risk Table for the potential NEA's The system is highly automated, and analyzes the asteroid catalog for potential problems. The catalog is continuously updated from Earth- and space-based observatories. Membership in the database is based on calculated probabilities of coming close to Earth. With better observations, some objects are removed from the database. There are "lost" objects on the database that have not been seen recently. There are two risk scores calculated, a general one (Torino Scale) and a more technical one (Palermo Scale).

The Cambridge, MA-based Minor Planet Center receives observation data from numerous sources around the globe. It calculates the orbital elements and risk scores, and updates the database accordingly. It has been operating since 1947

Orbital parameters are updated as more observations are acquired, and the Center does an orbital propagation 100 years into the future. If the object is going to get close enough to Earth to cause concern, it is added to the Earth Close Approach Table.

Sometimes, a minor planet falls out of view for a while, and is considered lost. This is when it is not seen where it was expected to be. Some of these are rediscovered later.

NASA's Planetary Defense Coordination Office

This organization is under NASA's Planetary Science Division. It is tasked with tracking and cataloging NEO's larger than 30 meters.

Asteroid impact avoidance

Well, we can't quite nudge the Earth to avoid a significant asteroid, but we might be able to nudge the asteroid. If its detected soon enough, we will have the time to implement a solution. It doesn't need to be moved much. For example, we might spray paint the surface, so it reflects more sunlight. This would actually slow it down, and change its trajectory. We might also nudge it with an ion propulsion engine, or attach a giant solar sail to slow it down. All of this has to be thought out and modeled very carefully so we solve the problem, not make it worse. What we don't want to do is detonate a nuclear weapon on it. We would get rid of the big chunk to have a million medium-sized chunks on the same trajectory. If we really have enough time, we can set up a spacecraft to just fly along side of the threat, and change its trajectory due to mutual gravity. We could paint it white or silver. Stay with me here. That would increase the reflectivity of the surface, making it more susceptible to radiation pressure from the sun. It has also been proposed to attach a large ballast mass, maybe another asteroid, with a tether. This changes the original asteroid's center of mass and hopefully its trajectory. Other options might include attaching a ion thruster using asteroid dust as the fuel. Just don't think about the nuclear option.

Non-Near Earth Issues

You know how much more complicated an accident is when you're away from home? Well, NASA's OSIRIS-REx mission was heading to an asteroid called Bennu, which is classified as a PHO. At a distance of 1 million kilometers, OSIRIS began keeping an eye out for debris in its path. These would include dust plumes and moon around asteroids. As of this writing, no dust plumes have been spotted. It has an impactor, and is taksed with returning some surface samples.

Later, the spacecraft will begin looking for mini-moons around Bennu, down around 10 centimeters. If it spots a moon, it will stop 25 or so miles away, and get more data on the moon's orbit.

If all goes well, it will continue its scan as it approaches the asteroid in 2019. If it sees dust plumes or mini-moons, so much the better, since little is know about these as well. Otherwise, Osiris will continue to execute its primary sample return mission. It started back to Earth in 2021, and should arrive in 2023, It is in extended mission. No rest for the weary, Osirix will drop off its sampes and head out to near-Earth asteroid Apophis, It will study the asteroid for 18 months, with no sample return. The asteroid will have its closest approach to Earth in 2029.

The author worked on this project briefly.

NASA Orbital Debris Program Office

The NASA Orbital Debris Program Office is hosted at the Johnson Space Center, in Houston, Texas. Under their guidance, orbital debris models are developed and tested. Some models are used to predict future orbital debris environments. Orbital debris is tracked optically and via ground based radar. Examining spacecraft parts that have been returned from orbit by Shuttle-based servicing missions is very useful. The Shuttles themselves provided good examples of debris hits. Models are validated by JSC's Hypervelocity Impact Technology Facility.

The Inter-agency Space Debris Coordination Committee consists of members of 10 space-faring countries, and ESA, and addresses the growing problem.

The U. S.'s Orbital Debris Mitigation Standard Practices addresses the debris problem. It defines mitigation standard practices for satellites and upper stages to control release of particles greater than 5 mm, that will remain on-orbit for 25 years or more. A part of this is to assess the mission in terms of limiting the probability for explosion. This applies to all onboard sources of stored energy. Spacecraft are required to be designed such that collisions with debris smaller than 1 cm will not cause loss of control. Lethal, non-trackable debris is defined as that less than 10 cm. The USAF's Space Surveillance Network can track items as small as 10 cm in LEO, or 1 meter in GEO.

End of life disposal is also addressed in three areas, atmospheric reentry, storage orbit, and retrieval. The reentry option assumes that the atmospheric drag will cause reentry within 25 years. This may involve the deployment of drag enhancement devices. Since even Cubesats are subject to this restriction, specific drag-enhancements for them have been developed. Reentry presents another problem, that of danger to humans and structures on the ground. The satellite must be re-entered in such a way that it enters the ocean in a remote location, with a 1 in 10,000 chance of human casualty. So far, so good.

A second option is maneuvering to a storage, or "graveyard" orbit. This doesn't solve the problem, it just postpones it. In the future, we can foresee an industry of spacecraft recycling in orbit.

Some of the Saturn upper stages on the lunar missions were deliberately placed into an orbit that would send them into the Sun. Between LEO and medium orbits, the spacecraft is supposed to be placed with a perigee above 2,000 km, and an apogee below 19,700 km (below synchronous orbit). For medium altitude orbits, up to GEO, the storage orbit is a perigee above 20,700 km, and a apogee above 35,300 km. This will be below synchronous altitudes. For geosynchronous altitude, the graveyard orbit is 300 km above the geosynchronous altitude. If the spacecraft is in a heliocentric orbit (around the Sun), it can also be maneuvered to collide with the Sun for disposal.

The third option is retrieval, which is difficult now with the retirement of the Space Shuttle. That also makes the option of on-orbit repair moot at the moment, but NASA is actively exploring in-orbit robotic repair at the Satellite Servicing Capabilities Office, at GSFC. Active removal involves more than technical issues. There are legal and ownership issues. On the high seas, it is the law that if you retrieve an un-crewed ship, it is yours. That's not true in space.

RemoveDebris

The Remov-Debris Project is addressing some possible solutions at the Surrey Space Centre in the U.K. The hardware went to the ISS, and was deployed by the crew. It comes with its own debris – several smallsats. These will be released, and Remov-Debris will attempt to net them. It's main approach is a harpoon. The third part of the technology demonstration is that it will be sent into the atmosphere to burn up, using a 10 square meter drag chute. As opposed to most missions, reentry and burning are the goals. The first test of this system seemed to have worked.

The Remove Debris mission was deployed from the ISS in mid-September, 2018, and worked as expected. The net deployed properly. The Cubesat target expanded a balloon to increase its effective size and thus drag. It is expected to reenter within a month. The next step is for the clean-up spacecraft to test out its harpoon with a flat target it will deploy, then deploy a sail, which will expedite reentry.

Airbus Industries is trying a variation of this approach for the elimination of the Envisat Earth Observation Platform, a defunct ESA Mission launched in 2002. It is somewhat embarrassing to have the world's largest environmental satellite cluttering up orbit. It masses some 8 metric tons. A harpoon perpetrator is fired with compressed gas, and easily penetrates the 3 cm composite honeycomb body panels. Then a set of spring loaded barbs extend, making the harpoon impossible to pull out. This harpoon is larger than the one on the Remove-Debris Project.

After the harpooned spacecraft is captured, the capturing satellite does a controlled reentry, and both spacecraft burn in the atmosphere.

WISE

The Wide-Field Infrared Survey Explorer launched in 2009. It operated until 2011, when it was put into hibernation mode, but was reactivated in 2013. It's accomplishments include the discovery of thousands of minor planets, and many star clusters. It was the first to see Earth's Trojan, named 210 TK7.

It provided observations of over 33,000 new asteroids, and 154,000 other solar system objects, 290 near-Earth asteroids and comets.

The mission was extended by one month in 2010 for the NeoWise searching for near Earth Objects. At this point the onboard cyrogen, used to cool two of the four detectors, had been depleted. It was successful, and the mission got extended by another 3 months. The spacecraft had characterized more than 150,000 minor planets, and discovered 35,000.

WISE completed a full scan of the asteroid belt, and was then put in hibernation, but was awakened now and then to check status. In August of 2013, it was recommissioned to search for near-Earth objects. It was also tasked to find asteroids that could be nudged into lunar orbit. This mission continued for 3 years, possibly affected by the Chelyabinsk meteor explosion, that had gone undetected until the explosion. The main telescope was cooled by having it stare into deep space for a while, before getting back to work again.

It was expected to discover 150 new NEO's and to do additional characteristics of some 2,000 known asteroids. It's record by May of 2018 was the discovery of 290 new NEO's,

NASA's Asteroid Redirect Mission

This project was proposed by NASA in 2013, with several modes. In one option, a robotic spacecraft would rendezvous and grapple an asteroid of around 4 meters. The target would be analyzed, then transported to a lunar orbit. An associated future mission would involve an Orion crewed vehicle visit. The program was not funded past the study phase. In one sense, this is like not buying a fire extinguisher, because you've never needed one.

The concept of a gravity tractor was proposed to nudge potentially hazardous asteroids off their course. No direct contact is required. The gravitational force between the asteroid and the tractor/spacecraft would be enough to modify the asteroid's orbit to a safer one, over a period of time. This takes a lot of patience, and assumes early detection. This approach is independent of the objects rotation. By tracking the tractor, we would know where the asteroid is at all times.

The change in velocity required to adjust the asteroid's trajectory can be rather small. For an example of a 100 meter object, only 1 centimeter/second velocity change would be sufficient. This would have to be applied over a period of 10 years, though. Again, early detection would be required.

There is a need for mature technology for rapid-response NEO reconnaissance, and deflection/destruction missions.

The latter could use nuclear devices (not a great idea), kinetic impactors, or gravity tractors. If we manage to deflect an asteroid, and it still impacts the Earth, are there international legal implications?

Technologies developed in this area will be applicable to asteroid mining, and the expanding use of space resources.

NEAR-Shoemaker

This mission was developed by the Applied Physics Laboratory of the Johns Hopkins University, and launched in 1996. It's mission was to get up close and personal with the NEO Eros for a year. It orbited the the asteroid, and made a successful touchdown on its surface in 1991. Eros was the second largest NEO known. NEAR provided data on chemical composition, and did a search for moons of Eros, but none were found.

EROS was the first spacecraft to soft land on an asteroid. It was shut down in 2001, after returning much valuable data.

Deep Impact Mission

Deep Impact was launched in 2005, heading to comet Tempel-1, to launch a penetrater to be able to study the composition. This was successful in creating an impact crater, and a large dust cloud. In extended mission, Deep Impact visited and imaged comet Hartley 2. It went on further missions to comets and asteroids, but communication with the spacecraft was lost in 2013.

Testing Earth's Planetary Defense

In October of 2017, NASA had an opportunity to test the international asteroid warning network. Asteroid 2012 TC4 went across Antarctica at around 26,000 feet altitude. The size of the asteroid was estimated to be around 30 meters. The event was broadcast across the globe, and numerous telescopes looked where the coordinates told them, and sent their data findings.

It was estimated that we are at least 10 years away from having a viable response to a hazard that a rock such as this could pose. At the moment, we can only document it, and wish real hard.

The Martian Planetary Defense Force

For a while in the late 1990's, it was postulated that the Martian Planetary Defense Force were very effective, as a large number of Mars-bound payloads failed. Actually, they weren't that good, but we Earthies were making mistakes big time.

The Mars Climate Orbiter crashed into the planet in 1999, instead of achieving orbit. Embarrassingly, the fault was one of units – a mix up in metric and English units in a propulsion parameter.

The Mars Rover Pathfinder, safely on the surface, suffered a series of resets, which turned out to be a software problem, and was fixed. This was a triumph of remote diagnostics and debugging.

The 2010 Russian Phobos Grunt mission never made it out of Earth orbit, and reentered and burned.

Launch Vehicles

You will note that the United States launches out over the Atlantic Ocean from Cape Kennedy, or over the Pacific from Vandenburg Air Force Base. Before a launch, mariners and air craft are warned of hazardous areas. It is common to drop the first stage booster in the ocean. Booster recovery and reuse is now routinely used by Space-X.

Sometimes, Nature itself causes a problem. In 1987, lightning struck a Atlas Centaur AC-67 rocket less than a minute after liftoff, causing it to explode.

DoD's Role, NORAD

Orbital Debris is tracked by radar. NORAD, the North American Aerospace Command, based in Colorado, tracks all detectable orbital entities, from large satellites to space junk, zombie-sats, and the larger pieces of debris, as well as near-Earth asteroids. The U. S. Space Surveillance Network can see objects 10 cm. or larger. U. S. Space Command deployed a new large telescope to improve their view of debris in 2011. This Space Surveillance Telescope is able to see debris at Geosynchronous altitudes. It has a 3.5 meter mirror, and has been in use since 2011 in Australia.

NORAD puts all this up on a website, in a standard format called the "two-line element" (TLE). This contains the Keplerian orbital elements, the set of data describing the orbit of anything around the Earth, for a given point in time (epoch). It is a legacy format form the 1960's, that still works. It includes two data items of 80 ASCII charters each (an IBM punch card format).

The MSX spacecraft, midcourse space experiment, was launched in 1996, with the primary mission of identifying and tracking ballistic missiles. It was built by the Johns Hopkins University Applied Physics Lab. After its cyrogenic coolant is depleted, it will
focus on objects in space, including keeping track f NEO's.

Other Spacefaring Nation's approaches and cooperation.

The International Asteroid Warning Network is a group of volunteer astronomers that observe and report on observations across the planet. We're all on this planet together.

Canada launched the NEOSSat, which is actually a microsat, It is described as suitcase-sized, but mounts a 15 centimeter telescope, and has 3-axis stabilization. It masses 74 kilograms. It is a joint project between the Canadian Space Agency, and the Defense Research and Development Canada, a lot like DARPA in the U. S.

It is a follow-on the Canada's MOST, a space telescope for astronomical observations.

NEOSSat takes exposures on the order of 100 seconds, and can detect sources down the magnitude 20. The mission was launched in 2013.

The spacecraft carries out surveillance of nearby space, with a mission of finding and tracking near Earth asteroids, those within Earth's orbit around the Sun. It also tracks some spacecraft in Earth's orbit as known objects, for calibration.

Darpa's Space Surveillance Telescope is located in Exmouth, Australia. It is a key part of the Australia-U.S. Space Situational Awareness Initiative. It has a view of the sky of the Southern Hemisphere.

The U. K. hosts the SpaceGuard Center. This term was used in Arthur C. Clarke's book series, *Rendezvous with*

Rama, concerning visits from a large automated spacecraft from outside our solar system. Dr. Clarke is considered the father of the Communications satellite.

The Spaceguard Foundation is an international association, founded right after a key conference in 1995. Several major objects, including the Chelyabinsk meteor which hit Russia, were not detected.

Japan has a (non-profit) private Spaceguard association, with an observatory at Bisei-cho, in the western part of the country. It keeps an eye out for orbital debris and NEO's. Significantly, it has an outreach program for students to learn astronomy, and the perils of space junk.

There was a Japanese mission to the asteroid Ryugu, discovered in 1999. The Hayabusa2 mission was launched late in 2014, and arrived at the minor body in 2018. It is tasked to collect samples, and return them to Earth by 2020.

The spacecraft includes 4 surface rovers. Just recently, the first two were deployed to the surface, followed by the joint French-German Mobile Asteroid Surface Scout. Battery life limited the mission to 16 hours of operation

Near Earth object deflection and disaster response is considered by the United Nations as part of its committee on Peaceful Uses of Outer Space. The U.N was assisted in addressing these issues by the B162 Foundation, and the Association of Space Explorers. One result of this is the International Asteroid Warning Network. Rogue asteroids are not impressed by national boundaries, and an impact event could effect multiple countries.

ESA has the NEOshield project, which is looking at technology and operational techniques to operate a spacecraft in close proximity to NEO's. Another goal is to refine NEO characterization. NEODyS is their online database of known NEOs.

Project Icarus

Icarus was a project in 1967 by MIT system engineering grad students to define a defense against asteroid 1566 Icarus, as a class project. Their proposed solution was a nuclear weapon.

Wrap-up

We know for a fact that a large asteroid is going to hit the Earth and do damage. It has happened before. What we don't know is when and where. We also have a few ideas on how to prevent that, but little has really been tested.

In the mean time, we observe the sky for potential problems, and try to categorize what we observe. It may take another near-extinction event to convince the non-scientists that its a good idea to spend some real money on this.

Glossary of Terms

AIAA – American Institute of Aeronautics and Astronautics.
ALA – asteroid laser ablation.
Albedo – reflectivity.
AoA – analysis of alternatives.
Aphelion – point in orbit farthest from Sun.
APL – Applied Physics Laboratory of the Johns Hopkins University.
Apogee – point in orbit farthest from the Earth
ARCM – Asteroid Retrieval Crewed Mission.
ARM – asteroid redirect mission.
ARRM – Asteroid Redirect Crewed Mission.
ARU – Asteroid retrieval and unilization mission (NASA)
ASAT – anti-satellite (weapon).
ASCII – American Standard Code for Information Interchange.
ASE – Association of Space Explorers.
Astrobleme – remains of an impact crater.
Asteroid - minor planet, in solar orbit.
Aten asteroid – Earth crossing bodies, 1,300 potentially hazardous objects.
Atira – class of asteroids within Earth's orbit.
ATLAS - Asteroid Terrestrial-impact Last Alert System.
    B612 Foundation – Formed to protect Earth from asteroid impacts.
BMD – Ballistic Missile Defense.
Bolide – Greek for missile. Very large impactor. Leaves a crater.

Cislunar – between the Earth and the moon.
CME – Coronal Mass Ejection.
CNSA – China National Space Administration.
CNEOS - Center for Near Earth Object Studies.
Copous – (U. N.) Committee on the Peaceful Uses of Outer Space.
CSA – Canadian Space Agency.
CTBTO - Comprehensive Test Ban Treaty Organization.
DAMien – Detecting and mitigating the Impact of Earth-bound Near Earth Objects.
DARPA – (U. S.) Defense Advanced Research Projects Agency.
DHS – (U.S.) Department of Homeland Security.
DLR – German Space Agency (Deutsches Zentrum für Luft- und Raumfahrt).
DoD – (U.S.) Department of Defense.
DOE – (U.S.) Department of Energy.
DRDC – Defense Research and Development Canada.
DRO – distance retrograde orbit.
EADP – (Danish) Emergency Asteroid Defense Project.
ECAT - Earth Close Approach Table.
EDEIS - Expert Database on Earth Impact Structures.
EID - Earth Impact Database.
EIEP – Earth Impact Effects Program.
ESA – European Space Agency.
EVA – extra-vehicular activity. outside the spacecraft.
Exeroid – object from outside of our solar system.
FEMA – (U. S.) Federal Emergency Management Agency,
GEO – geosynchronous earth orbit.
GPU – graphics processing unit.

Graveyard orbit – a place to put end-of-life satellites.
GSFC – Goddard Space Flight Center, Greenbelt, Maryland. NASA Center for unmanned spacecraft near Earth.
HAIV - Hypervelocity Asteroid Intercept Vehicle. (concept)
Hammer - Hypervelocity Asteroid Mitigation Mission for Emergency Response.
HEOSS - High Earth Orbit Space Surveillance.
IADC – Interagency Space Debris Coordination Committee.
IAU – International Astronomical Union.
IAWN – International Asteroid Warning Network.
IBM – International Business Machines, computer manufacturer.
IFSG - Impact Field Studies Group.
IIAWN - International Asteroid Warning Network.
ISS – International Space Station.
IWG – Interagency Working Group
JAXA - Japan Aerospace Exploration Agency.
JHU – Johns Hopkins University.
JSC – NASA Johnson Space Center, Texas.
JSOC – U. S. Joint Space Operations Center
JPL – Jet Propulsion Lab.
KT – kilo ton.
LEM – lunar excursion module (crewed).
LEO – Low Earth orbit.
LLNL - Lawrence Livermore National Laboratory.
LSST - Large Synoptic Survey Telescope.
MBPL – Minor Body Priority List.
Magnitude - in astronomy, a log scale of brightness.

MASCOT - Mobile Asteroid Surface Scout.

Meteor - glowing meteoroid, or comet, or asteroid passing through the atmosphere.

Meteor Crater – Arizona. Also known as Barringer Crater.

Meteoroid – rocky or metallic body in space, smaller than 1 meter.

Meteorite – solid debris, enters atmosphere and reaches the surface.

MMOD - micrometeoroid and orbital debris.

MOID - Minimum Orbit Intersection Distance.

MOST - Microvariability and Oscillations of Stars telescope (Canada).

MPC – Minor Planet Center, Cambridge, Ma.

MPC's – minor planet circulars, from MPC.

MPO – Minor planets and comets orbit supplement, from MPC.

MPS – Minor Planet Supplement, from MPC.

MSX – midcourse space experiment.

MT – mega-ton.

NASA – National Aeronautics and Space Administration

NEA – Near Earth Asteroid.

NEAR Shoemaker - Near Earth Asteroid Rendezvous – Shoemaker.

NEAT - Near-Earth Asteroid Tracking.

NEC – Near Earth Comet.

NIST – National Institute of Standards and Technology.

NEO – Near Earth Object, any significant object within 30 million miles of the home planet.

NEODyS – ESA online database of known NEOs.

NEOP – Near Earth Object Program (JPL).

NEOSSat, Near Earth Object Surveillance Satellite (Canada)
NEOWISE - Near Earth Object WISE (Wide-field Infrared Survey Explorer).
NESS - Near Earth Space Surveillance.
NGO – non-government organizations
NOAA – (U.S.) National Oceanic and Atmospheric Administration.
NORAD – North American Air Defense command (USAF).
NSF – (U.S.) National Science Foundation.
NSTC – (U.S.) National Science Technology Council.
ODPO – (NASA) Orbital Debris Program Office.
ODQN – Orbital Debris Quarterly News, from NASA, JSC.
OSIRIS-REx- (Origins, Spectral Interpretation, Resource Identification, Security-Regolith Explore (spacecraft).
OSTP – (U.S.) Office of Science and Technology Policy.
PA&E – Program Analysis and Evaluation.
Palermo scale – rates potential of NEO impact.
Pan-STARRS Panoramic Survey Telescope & Rapid Response System.
PDCO – (NASA) Planetary Defense Coordination Office.
Perihelion – point in orbit closest to the Sun.
Petabyte – 10^{15} bytes.
Perigee – point in orbit closest to the Earth.
PHA – potentially hazardous object.
PHO – potentially hazardous object.
PI – Principal Investigator.

PRA – Probabilistic Risk Assessment.
Roche limit – the distance in which an orbiting body will be torn apart by tidal forces of the primary.
ROS – Russian Orbital Segment of the ISS.
SAO – Smithsonian Astrophysics Observatory.
Semi-major axis – length of the shorter diameter of an ellipse.
SEP – solar electric propulsion
SI – System Internationale (metric).
SMPAG – Space Mission Planning Advisory Group.
SNAP - Systems for Nuclear, Auxiliary Power.
Socrates – Satellite Orbit Conjunction Reports Assessing Threatening Encounters in Space.
SST – space surveillance telescope (DARPA).
Super bolide - bolide that reaches magnitude -17 or brighter.
TAGSAM - Touch-And-Go Sample Acquisition Mechanism.
TECA – Table of Asteroids (Next) Closest Approaches to the Earth.
TNT – Trinitrotoluene, explosive material.
Torino Scale – rates the impact hazard associated with NEO's.
USAF – United States Air Force.
USGS - United States Geological Survey.
VA – virtual asteroid.
WGNEO - Working Group on Near-Earth Objects
WMOPS - WISE Moving Object Processing Software
Zombie-Sat – out-of-control non-responsive satellite posing a danger to other spacecraft.
ZTF - Zwicky Transient Facility.

Bibliography

Abell, P. A. *Missions to Near-Earth Asteroids: Implications for Exploration, Science, Resource Utilization, and Planetary Defense*, 2012, ASIN-B01I8TR1HO.

Abel, P. A. *NASA's Asteroid Redirect Mission: A Robotic Boulder Capture Option for Science, Human Exploration, Resource Utilization, and Planetary Defense*, 2014, ASIN-B01FKLMC9A.

Adams, R. B. *Planetary Defense: Options for Deflection of Near Earth Objects*, 2013,R. B. Adams NASA Technical Reports Server, 2013, ISBN-1287284485.

Bell, Larry D. *Planetary Asteroid Defense Study: Assessing and Responding to the Natural Space Debris Threat* , 2012, ISBN-1288329687.

Blaauw, R. C. *When the Sky Falls NASA's Response to Bright Bolide Events Over Continental USA*, 2015, ASIN-B01C42OS7C.

Blaauw, R. C. *NASA's Meteoroid Environments Office's Response to Three Significant Bolide Events Over North America*, 2015, ASIN-B01C42O00W.

Clarke, Arthur C. *Rendezvous with Rama*, 2012, ASIN-B01IA89F9U. (Novel)

Cristea, Emil M. "Planetary Defense," 2016, U. S. Air Force Air Command and Staff College, Air University, Avail: http://www.dtic.mil/dtic/tr/fulltext/u2/1031581.pdf

Glasstone, Samuel, *Effects of Nuclear Weapons*, 2013, ISBN-1258789531.

Hughes, Gary B. *Planetary Defense and Space Environment Applications* (Proceedings of SPIE), 2017, ISBN-1510603530.

JPL, "Twenty years of Planetary Defense," 2018, avail: https://phys.org/news/2018-07-twenty-years-planetary-defense.html

Kleiman, Louis A., *Project Icarus: an MIT Student Project in Systems Engineering,* MIT Press, 1968, ISBN-0-262-63068-0.

O'Keefe, J. D.; Ahrens, Thomas J. "Impact Mechanics of the Cretaceous–Tertiary Extinction Bolide," 1982, Nature 298, pp 123–127. avail: https://www.nature.com/articles/298123a0

Pelton, Joseph N. ; Allahdadi, Firooz *Handbook of Cosmic Hazards and Planetary Defense,* 2015, Springer, ISBN-3319039512.

Pournelle, Jerry; Niven, Larry *Lucifer's Hammer*, 1985, ISBN-0449208133.

Nahra, Henry ."Effect of Micro Meteoroid and Space Debris Impacts on the Space Station Freedom Solar Array Surfaces". NASA, 1989, avail: https://ntrs.nasa.gov/archive/nasa/casi.ntrs.nasa.gov/19890016664.pdf

Nigi, Rosario *Planetary Defense. Department of Defense Cost for the Detection, Exploration, and Rendezvous Mission of Near-Earth Objects*, 1997.

Sagan, Carl; Druyan, Ann *Pale Blue Dot: A Vision of the Human Future in Space,* 2011, ASIN-B004W0I3LW.

Stakem, Patrick H. *The Saturn Rocket and the Pegasus Missions, 1965,* 2013, PRRB Publishing, ISBN-1520209916. .

Symons, G. J. *On the Detonating Bolide of November 20th, 1887,* Proceedings of the Royal Society of London, 1887, ASIN-B01AGVV5LU.

Taylor, Travis S.; Boan, Bob *An Introduction to Planetary Defense: A Study of Modern Warfare Applied to Extra-Terrestrial Invasion*, 2006, ISBN-1581124473.

U. S. Air Force Air Command and Staff College, *Should the USAF be Involved in Planetary Defense, 2014,* ISBN-1499776659.

U.S. Air Force Research Laboratory, *On Near Earth Objects Threat Mitigation* (Planetary Defense), 2014,

ISBN-1499790694.

Wells, H. G. *The Star*, ISBN-1535420563. (Novel)

Resources

- Lawrence Livermore National Laboratory Planetary Defense Workshop, May 22-26, 1995, ASIN-B015LDPEIS.
- NASA Planetary Defense Office,
- https://www.nasa.gov/planetarydefense
- http://www.au.af.mil/au/afri/aspj/apjinternational/apj-s/2009/1tri09/franceeng.htm
- http://www.citizensinspace.org/category/past-and-future/planetary-defense/
- https://bigthink.com/philip-perry/nasa-just-tested-earths-planetary-defense-system
- Earth Impact Database: http://www.passc.net/EarthImpactDatabase/
- Earth Impact Effects Program: https://impact.ese.ic.ac.uk/ImpactEarth/ImpactEffects/
- https://cneos.jpl.nasa.gov/fireballs/
- https://www.whitehouse.gov/wp-content/uploads/2018/06/National-Near-Earth-Object-Preparedness-Strategy-and-Action-Plan-23-pages-1MB.pdf
- https://meteoritical.org/
- IAU Minor Planet Center, list of Potentially Hazardous Asteroids, https://www.minorplanetcenter.net/iau/lists/t_phas.html
- NASA/JPL/Center for Near Earth Object Studies, Earth Impact Monitoring - https://cneos.jpl.nasa.gov/sentry/

- Report of the Task Force on potentially hazardous Near Earth Objects – avail: http://space.nss.org/media/2000-Report-Of-The-Task-Force-On-Potentially-Hazardous-Near-Earth-Objects-UK.pdf
- http://www.passc.net/EarthImpactDatabase/index.html
- https://cneos.jpl.nasa.gov/
- https://www.nasa.gov/planetarydefense/overview
- http://www.bbc.com/news/science-environment-24839601
- http://www.spacesafetymagazine.com/space-hazards/asteroid-hitting-earth/Center for NEO Studies, https://cneos.jpl.nasa.gov/
- https://bigthink.com/philip-perry/nasa-just-tested-earths-planetary-defense-system
- https://www.nasa.gov/audience/forstudents/k-4/more_to_explore/Asteroids-Comets-Meteorites.html
- http://www.passc.net/EarthImpactDatabase/
- wikipedia, various.

If you enjoyed this book, you might also be interested in some of these.

Stakem, Patrick H. *Floating Point Computation*, 2013, PRRB Publishing, ISBN-152021619X.

Stakem, Patrick H. *Architecture of Massively Parallel Microprocessor Systems*, 2011, PRRB Publishing, ISBN-1520250061.

Stakem, Patrick H. *Multicore Computer Architecture*, 2014, PRRB Publishing, ISBN-1520241372.

Stakem, Patrick H. *Personal Robots*, 2014, PRRB Publishing, ISBN-1520216254.

Stakem, Patrick H. *RISC Microprocessors, History and Overview,* 2013, PRRB Publishing, ISBN-1520216289.

Stakem, Patrick H. *Robots and Telerobots in Space Application*s, 2011, PRRB Publishing, ISBN-1520210361.

Stakem, Patrick H. *The Saturn Rocket and the Pegasus Missions, 1965,* 2013, PRRB Publishing, ISBN-1520209916.

Stakem, Patrick H. *Visiting the NASA Centers, and Locations of Historic Rockets & Spacecraft,* 2017, PRRB Publishing, ISBN-1549651205.

Stakem, Patrick H. *Microprocessors in Space*, 2011, PRRB Publishing, ISBN-1520216343.

Stakem, Patrick H. Computer *Virtualization and the Cloud*, 2013, PRRB Publishing, ISBN-152021636X.

Stakem, Patrick H. *What's the Worst That Could Happen? Bad Assumptions, Ignorance, Failures and Screw-ups in Engineering Projects, 2014*, PRRB Publishing, ISBN-1520207166.

Stakem, Patrick H. *Computer Architecture & Programming of the Intel x86 Family, 2013*, PRRB Publishing, ISBN-1520263724.

Stakem, Patrick H. *The Hardware and Software Architecture of the Transputer*, 2011,PRRB Publishing, ISBN-152020681X.

Stakem, Patrick H. *Mainframes, Computing on Big Iron*, 2015, PRRB Publishing, ISBN- 1520216459.

Stakem, Patrick H. *Spacecraft Control Centers*, 2015, PRRB Publishing, ISBN-1520200617.

Stakem, Patrick H. *Embedded in Space,* 2015, PRRB Publishing, ISBN-1520215916.

Stakem, Patrick H. *A Practitioner's Guide to RISC Microprocessor Architecture*, Wiley-Interscience, 1996,

ISBN-0471130184.

Stakem, Patrick H. *Cubesat Engineering*, PRRB Publishing, 2017, ISBN-1520754019.

Stakem, Patrick H. *Cubesat Operations*, PRRB Publishing, 2017, ISBN-152076717X.

Stakem, Patrick H. *Interplanetary Cubesats*, PRRB Publishing, 2017, ISBN-1520766173 .

Stakem, Patrick H. Cubesat Constellations, Clusters, and Swarms, Stakem, PRRB Publishing, 2017, ISBN-1520767544.

Stakem, Patrick H. *Graphics Processing Units, an overview*, 2017, PRRB Publishing, ISBN-1520879695.

Stakem, Patrick H. *Intel Embedded and the Arduino-101, 2017,* PRRB Publishing, ISBN-1520879296.

Stakem, Patrick H. *Orbital Debris, the problem and the mitigation*, 2018, PRRB Publishing, ISBN-*1980466483*.

Stakem, Patrick H. *Manufacturing in Space*, 2018, PRRB Publishing, ISBN-1977076041.

Stakem, Patrick H. *NASA's Ships and Planes*, 2018, PRRB Publishing, ISBN-1977076823.

Stakem, Patrick H. *Space Tourism*, 2018, PRRB

Publishing, ISBN-1977073506.

Stakem, Patrick H. *STEM – Data Storage and Communications*, 2018, PRRB Publishing, ISBN-1977073115.

Stakem, Patrick H. *In-Space Robotic Repair and Servicing*, 2018, PRRB Publishing, ISBN-1980478236.

Stakem, Patrick H. *Introducing Weather in the pre-K to 12 Curricula, A Resource Guide for Educators*, 2017, PRRB Publishing, ISBN-1980638241.

Stakem, Patrick H. *Introducing Astronomy in the pre-K to 12 Curricula, A Resource Guide for Educators*, 2017, PRRB Publishing, ISBN-198104065X.
Also available in a Brazilian Portuguese edition, ISBN-1983106127.

Stakem, Patrick H. *Deep Space Gateways, the Moon and Beyond*, 2017, PRRB Publishing, ISBN-1973465701.

Stakem, Patrick H. *Exploration of the Gas Giants, Space Missions to Jupiter, Saturn, Uranus, and Neptune*, PRRB Publishing, 2018, ISBN-9781717814500.

Stakem, Patrick H. *Crewed Spacecraft*, 2017, PRRB Publishing, ISBN-1549992406.

Stakem, Patrick H. *Rocketplanes to Space*, 2017, PRRB

Publishing, ISBN-1549992589.

Stakem, Patrick H. *Crewed Space Stations,* 2017, PRRB Publishing, ISBN-1549992228.

Stakem, Patrick H. *Enviro-bots for STEM: Using Robotics in the pre-K to 12 Curricula, A Resource Guide for Educators,* 2017, PRRB Publishing, ISBN-1549656619.

Stakem, Patrick H. *STEM-Sat, Using Cubesats in the pre-K to 12 Curricula, A Resource Guide for Educators*, 2017, ISBN-1549656376.

Stakem, Patrick H. *Lunar Orbital Platform-Gateway*, 2018, PRRB Publishing, ISBN-1980498628.

Stakem, Patrick H. *Embedded GPU's*, 2018, PRRB Publishing, ISBN- 1980476497.

Stakem, Patrick H. *Mobile Cloud Robotics*, 2018, PRRB Publishing, ISBN- 1980488088.

Stakem, Patrick H. *Extreme Environment Embedded Systems,* 2017, PRRB Publishing, ISBN-1520215967.

Stakem, Patrick H. *What's the Worst, Volume-2*, 2018, ISBN-1981005579.

Stakem, Patrick H., *Spaceports*, 2018, ISBN-1981022287.

Stakem, Patrick H., *Space Launch Vehicles*, 2018, ISBN-1983071773.

Stakem, Patrick H. *Mars*, 2018, ISBN-1983116902.

Stakem, Patrick H. *X-86, 40th Anniversary ed*, 2018, ISBN-1983189405.

Stakem, Patrick H. *Lunar Orbital Platform-Gateway*, 2018, PRRB Publishing, ISBN-1980498628.

Stakem, Patrick H. *Space Weather*, 2018, ISBN-1723904023.

Stakem, Patrick H. *STEM-Engineering Process*, 2017, ISBN-1983196517.

Stakem, Patrick H. *Space Telescopes,* 2018, PRRB Publishing, ISBN-1728728568.

Stakem, Patrick H. *Exoplanets*, 2018, PRRB Publishing, ISBN-9781731385055.

Stakem, Patrick H. *Planetary Defense*, 2018, PRRB Publishing, ISBN-9781731001207.

Patrick H. Stakem *Exploration of the Asteroid Belt*, 2018, PRRB Publishing, ISBN-1731049846.

Patrick H. Stakem *Terraforming*, 2018, PRRB

Publishing, ISBN-1790308100.

Patrick H. Stakem, *Martian Railroad,* 2019, PRRB Publishing, ISBN-1794488243.

Patrick H. Stakem, *Exoplanets,* 2019, PRRB Publishing, ISBN-1731385056.

Patrick H. Stakem, *Exploiting the Moon,* 2019, PRRB Publishing, ISBN-1091057850.

Patrick H. Stakem, *RISC-V, an Open Source Solution for Space Flight Computers,* 2019, PRRB Publishing, ISBN-1796434388.

Patrick H. Stakem, *Arm in Space*, 2019, PRRB Publishing, ISBN-9781099789137.

Patrick H. Stakem, *Extraterrestrial Life*, 2019, PRRB Publishing, ISBN-978-1072072188.

Patrick H. Stakem, *Space Command*, 2019, PRRB Publishing, ISBN-978-1693005398.

CubeRovers, A Synergy of Technologys, 2020, PRRB Publishing, ISBN-979-8651773138.

Robotic Exploration of the Icy moons of the Gas Giants. 2020, PRRB Publishing, ISBN- 979-8621431006

Hacking Cubesats, 2020, PRRB Publishing, ISBN-979-

8623458964.

History & Future of Cubesats, PRRB Publishing, ISBN-979-8649179386.

Hacking Cubesats, Cybersecurity in Space, 2020, PRRB Publishing, ISBN-979-8623458964.

Powerships, Powerbarges, Floating Wind Farms: electricity when and where you need it, 2021, PRRB Publishing, ISBN-979-8716199477.

Hospital Ships, Trains, and Aircraft, 2020, PRRB Publishing, ISBN-979-8642944349.

CubeRovers, a Synergy of Technologys, 2020, ISBN-979-8651773138

Exploration of Lunar & Martian Lava Tubes by Cube-X, ISBN-979-8621435325.

Robotic Exploration of the Icy moons of the Gas Giants, ISBN- 979-8621431006.

History & Future of Cubesats, ISBN-978-1986536356.

Robotic Exploration of the Icy Moons of the Ice Giants, by Swarms of Cubesats, ISBN-979-8621431006.

Swarm Robotics, ISBN-979-8534505948.

Introduction to Electric Power Systems, ISBN-979-8519208727.

Centros de Control: Operaciones en Satélites del Estándar CubeSat (Spanish Edition), 2021, ISBN-979-8510113068.

Exploration of Venus, 2022, ISBN-979-8484416110.

Patrick H. Stakem, *The Search for Extraterrestial Life,* 2019, PRRB Publishing, ISBN-1072072181.

The Artemis Missions, Return to the Moon, and on to Mars, 2021, ISBN-979-8490532361.

James Webb Space Telescope. A New Era in Astronomy, 2021, ISBN-979-8773857969.

www.ingramcontent.com/pod-product-compliance
Lightning Source LLC
Chambersburg PA
CBHW030502220526
45464CB00006B/2626